Sea Anemones
Science Under The Sea

Lynn M. Stone

Rourke

Publishing LLC

Vero Beach, Florida 32964

© 2003 Rourke Publishing LLC
www.rourkepublishing.com

PHOTO CREDITS: Cover, title page, p. 7, 16, 20 © Lynn M. Stone; p. 4, 12 © James H. Carmichael; p. 8, 10, 13, 15, 19 © Brandon Cole.

Cover Photo: *A sea anemone's stinging tentacles*

EDITOR: Frank Sloan

COVER DESIGN: Nicola Stratford

Library of Congress Cataloging-in-Publication Data

Stone, Lynn M.
 Sea anemones / Lynn M. Stone.
     p. cm. — (Science under the sea)
Summary: Describes the physical characteristics, behavior, and habitat of these flowerlike marine invertebrates.
Includes bibliographical references (p.   )
 ISBN 1-58952-317-2 (hardcover)
 1. Sea anemones—Juvenile literature. [1. Sea anemones.] I. Title.
 QL377.C7 S86 2002
 593.6—dc21
                                                      2002005129

Printed in the USA

CG/CG

Table of Contents

A Sea Flower

Discover a sea **anemone** and you'll think you've found a flower. Sea anemones are even named after a flower! But a sea anemone is a **marine** animal. It lives in the world's oceans, from the Arctic to the Antarctic.

There are nearly a thousand known **species** of sea anemones. Sea anemones are common in the shallow **tide pools** along the rocky coasts of North America. They live in many different types of ocean homes, even in deep sea canyons.

Delicate sea anemones look like undersea flowers.

Surrounded by Tentacles

Like flowers, sea anemones have a circular center. A flower's center is surrounded by petals. The sea anemone's center is surrounded by **tentacles**.

Sea anemone tentacles look and sway like pointed leaves. The number of tentacles changes with the kind of anemone. Some sea anemones have dozens of tentacles.

These sea anemones live in shallow water along the Oregon coast.

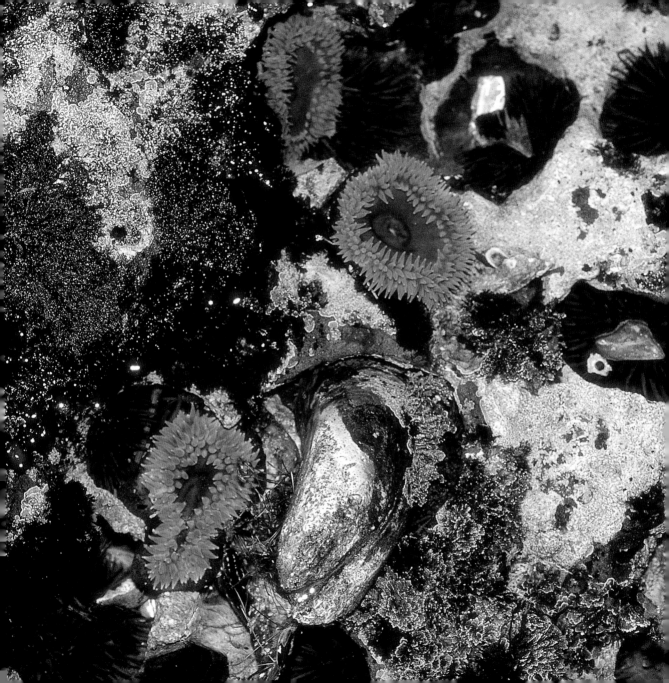

Animals without Backbones

Sea anemones are **invertebrates**. Basically, invertebrates are small animals without backbones, such as spiders, worms, insects, and snails. The sea anemones' family of invertebrates includes such animals as corals and jellyfish.

Tentacles of the crimson sea anemone wave like leaves of red grass.

The Sea Anemone's Body

A sea anemone's body is shaped like a long bag. The open top of the "bag" is the anemone's mouth.

A sea anemone brings food through its mouth into its hollow body. The animal returns food waste and even baby anemones out through the mouth.

Tentacles surround the mouth of a purple-tip sea anemone.

*The clownfish lives among stinging anemone tentacles, but the fish is
protected by a coat of slime called mucus.*

Baby sea anemones crawl out of their mother's mouth and onto her body, where they'll live until they grow up.

Living in One Place

The bottom side of the anemone "bag" is muscle. Sea anemones can creep slowly on their bottom sides. A few kinds of sea anemones can swim. But most sea anemones spend their lives in one place. They attach themselves to a rock or other hard surface.

Sea anemones can be as small as thimbles or up to 3 feet (.91 meters) across.

Finding a parking place can be tough for animals that don't travel, so strawberry sea anemones have crowded onto the shell of a giant mussel.

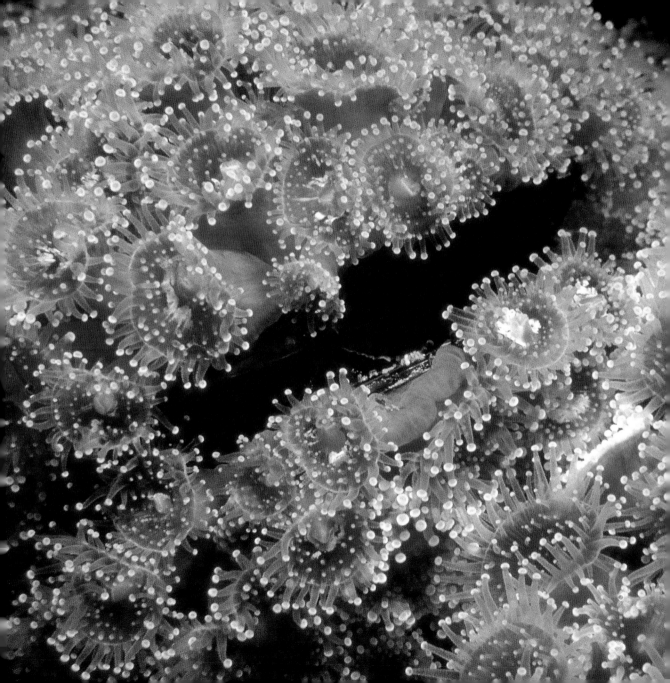

Predator and Prey

Don't be fooled by the sea anemone's beauty. Sea anemones are **predators**. They're not exactly in the class of lions and leopards. However, they do live on the flesh of other animals.

A sea anemone can't see or chase its prey. **Prey** comes to the sea anemone. When a small marine animal brushes against the sea anemone, the flowery tentacles become weapons.

A giant green sea anemone tries to eat a sea star.

Poisonous Darts

Each tentacle releases tiny poisonous darts. These darts are too small for us to see. But they stick to, wrap around, and poison small fish and other prey. The tentacles then move prey into the sea anemone's mouth.

A sea anemone's stinging tentacles slowly unfold from within its body.

Catching Prey from All Directions

With its wheel shape of tentacles, a sea anemone can catch prey from any direction. The circle of tentacles also gives it protection from all directions.

If disturbed, sea anemones pull in their tentacles and close up. Then they look more like fleshy lumps than flower animals!

Looking like donuts, sea anemones close up as the ocean tide lowers the water level above them.

Anemonefish

A few kinds of fish are not hurt by the anemone's dangerous tentacles. These are the colorful anemonefish, or clownfish. They live among ten species of anemones in warm parts of the Indian and Pacific oceans.

Bigger fish don't attack the anemonefish because they fear the anemone tentacles. The anemones, in turn, get food scraps that are left by the clownfish.

Glossary

anemone (an EM on ee) — a showy flower after which the sea anemone is named

invertebrates (in VUR tuh brates) — any of the boneless animals, such as worms, insects, lobsters, and snails

marine (meh REEN) — from or having to do with the sea

predators (PRED eh torz) — animals that hunt other animals for food

prey (PRAY) — an animal hunted by other animals

species (SPEE sheez) — within a group of closely related animals, one certain kind, such as a green sea anemone

tentacles (TEN tuh kelz) — soft, stalk-like structures used by sea anemones to grasp, cripple, and kill prey

tide pool (TIED POOL) —a shallow, seaside pool of ocean water that remains even when the ocean is in a state of low tide

Index

Further Reading

Frahm, Randy. *Oceans: Lifeblood of the Earth.* Creative Education, 2001
Schaefer, Lola M. *Sea Anemones.* Heinemann Library, 2002

Websites To Visit

Corals and Anemones: http://www.seasky.org/reeflife/sea2b.html
Sea Anemones: http://www.cyhaus.com/marine/anemone.htm

About The Author

Lynn Stone is the author of more than 400 children's nonfiction books. He is a talented natural history photographer as well. Lynn, a former teacher, travels worldwide to photograph wildlife in its natural habitat.